Letts

KS1
Success

Revision Guide

Lynn Huggins-Cooper

Science
SATs

Contents

Plants and animals

Humans

Materials

Physical processes

National Test practice

Answers and glossary

Plants

The parts of a plant

Have you ever picked a flower from the garden? Flowers are the parts of a plant that we like to look at and smell, but plants have lots of other important parts, too!

Plants need sunshine and water to grow

If I were a plant, I'd be a sunflower, as they're tall and strong.

I think you're more like a little weed!

The parts and their jobs

There are four important parts of a plant. Each part has a different job.

Leaf

The leaf uses sunshine to make food. Plants need sunshine and water to grow.

Flower

The flower often smells good and is brightly coloured, to attract insects. Insects help the plant to make seeds.

Roots

The roots help the plant to hold itself in the soil.

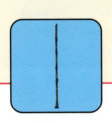

Stem

The stem carries water and goodness from the soil to all parts of the plant.

Top Tip

A stem is a bit like a straw. Suck water up through a straw to see how a stem works!

Have a go...

Dig up a weed, such as a dandelion. Describe to a grown-up what each part does.

Key words

flower	leaf
stem	roots

Quick Test

1. What job does a leaf do?
2. Why do plants have roots?
3. How is the stem like a straw?
4. Why are flowers often brightly coloured and nice-smelling?
5. What does a plant need in order to grow?

The five senses

Human senses

Humans have five **senses**.

We:

smell with our nose

taste with our tongue

touch with our fingers

hear with our ears

see with our eyes

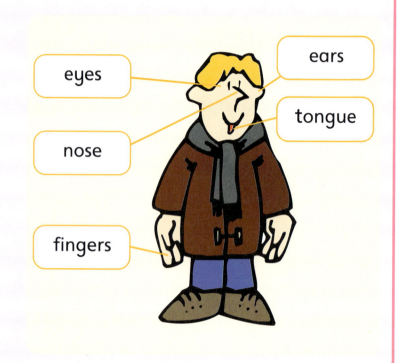

eyes

ears

tongue

nose

fingers

Use your senses!

We use our senses all the time. They **help us to find out things about the world** and how things work.

Look at your tongue in the mirror. Can you see some tiny bumps? They are called **taste buds** and they help you to taste your food.

Have you ever tasted something really sour? Think of sucking a lemon. Does it make your mouth water?

Top Tip

Taste a slice of lemon or lime. Can you feel your mouth water? What happens to your face?

Smelling and tasting

Smelling and tasting go together. Have you noticed that when you have a cold you cannot taste things very well? That is because you cannot smell them!

Your nose and mouth are joined inside by special tubes. Your nose gets blocked when you have a cold, so you cannot taste things properly.

If I had no nose, how would I smell?

You'd still smell horrible!

Have a go...

Go on a 'senses walk' with a grown-up. As you walk around your neighbourhood, think about the way your senses help you to understand the world. What senses did you use?

Key words

senses

taste buds

Quick Test

1. What are your five senses?
2. What are the bumps on your tongue called?
3. Why can't you taste things when you have a cold?
4. What are your nose and mouth joined together by?
5. Why do we need senses?

Recycling and reusing

Litter

Have you ever seen rubbish lying about? It looks horrible! **Litter is things that have been dropped on the floor, like old bags, packets and bottles.**

Litter can be dangerous, if you are an animal. Tiny animals such as mice can get stuck in bottles. Hedgehogs get tangled in the plastic that holds drink cans together. Foxes cut themselves on sharp cans and broken glass.

This is why it is very important that we do not drop litter!

Recycling is a great way to get new things from old junk!

Can I recycle you and get a new brother, then?

Caring for our world

We can help to **take care of our world** in lots of ways.

We can:

- Make sure we **take litter home** and put it in the bin.

- **Reuse** things like carrier bags, so there is less rubbish to throw away. We can also have fun making models from junk!

- Send bottles, cans, papers and plastics to be **recycled** and made into new things.

- Tell friends about what happens to wildlife when we drop rubbish.

Top Tip

Paint or cover a box to look like a big monster. Stick paper arms and legs on it and big wobbly eyes. Then ask a grown-up to help you to cut out a big mouth and you can use the monster as a 'garbage gobbler' – much more fun than a bin!

Have a go...

Design a poster on your computer to tell people not to drop litter. Use clip art and interesting fonts to make people take notice!

Key words

litter   recycled

reuse

Quick Test

1. What is litter?

2. How can sharp cans and glass be dangerous to animals?

3. How can we reuse things?

4. Why are dumped bottles dangerous to tiny creatures?

5. What word describes how materials can be made into new things?

When things are alive

Help the alien!

This alien has come to earth on a school trip to find out about our planet and the things that live here. He needs to send a message back to the planet Zork to tell his teacher about living things on earth.

He needs to let her know how she can tell whether things are **alive** or not.

Look at what he has written on the next page.

Well, we know Sam is alive – because he eats lots!

And we know Mel is alive, because she keeps growing out of her clothes!

What do all living things do?

I have noticed seven things that all living things do.

They all:

- feed
- move
- feel things
- breathe
- get rid of waste
- grow and change
- have babies

I was a bit confused at first, because I thought that **plants were alive**, but could not see them breathing. They do not have babies in the same way as cats and rabbits either. However, they do **need air to live** and they do **make more new plants**, so they must be alive!

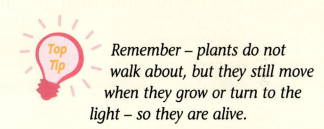

Top Tip

Remember – plants do not walk about, but they still move when they grow or turn to the light – so they are alive.

Have a go...

See if you can remember all seven things that show that something is alive. Draw a picture of a person and label it with your seven points.

Key words

alive

Quick Test

Which of the things in this list are alive? How do you know?

pebble   dandelion   tiger

river   oak tree   guinea pig

Sorting

Same or different?

When we need to sort a collection of things, we can look for **what is the same** and **what is different** about the things we need to sort.

Look at this group below. We can sort it into **sets** in two ways.

We could put the ladybird and earwig in one set, because they are both **insects**.

The cherries and flower could be in another set, because they are not insects.

The ladybird, flower and cherries could be in one set, because they are all red.

The earwig could be in another set, because it is not red.

Top Tip

Don't forget – when you have to sort things into sets, look for what is the same and what is different about them.

Sam and I are the same, because we're both children. But we're different, because he's a boy and I'm a girl.

And I'm cool and you're not!

Sorting animals into sets

We can sort animals into sets by looking at ways they are the same and ways they are different.

Look at this group:

crab   rabbit   cat   tortoise

The rabbit and the cat are **both furry**, so they could be in one set, and the tortoise and crab **both have shells**, so they could be in another set.

rabbit   cat   tortoise   crab

Have a go...

Sort out a pile of toys by looking for things that are the same and things that are different.

Key words

sets   insects

Quick Test

Look at the animals in this group. How could you sort them into two sets?

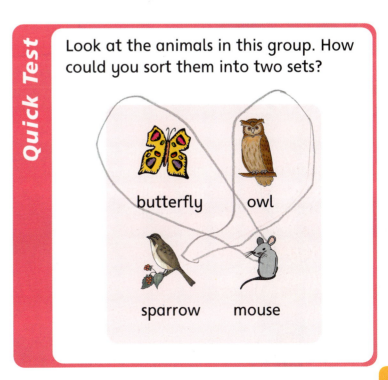

butterfly   owl

sparrow   mouse

Test your skills

How plant leaves get water

How does water get to the leaves of a plant?

You need:

- Celery sticks

- Old plastic container

- Food colouring – blue, red or orange are good colours to use

- Water

- Spoon

- Newspaper to cover surfaces

- Old shirt or apron

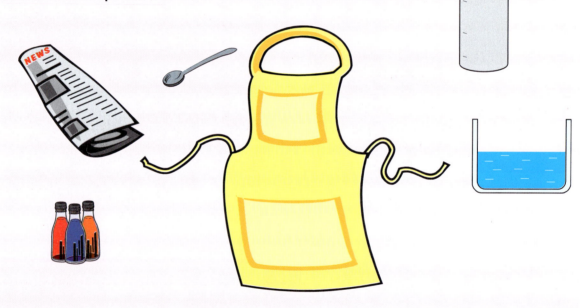

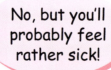
If I drink too much blackcurrant juice, will I turn purple?

No, but you'll probably feel rather sick!

What to do

1 Put a little water in your container.

2 Add 12 drops of food colouring – do not drop it on your clothes, because it will stain!

3 Stir the water to mix in the colouring. It should be a dark colour.

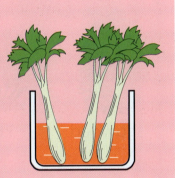

4 Carefully cut three stalks of celery and stand them in the coloured water.

5 Leave the celery in the water for a day. Check it now and again. What do you think will happen? After a day, cut the celery in half. You will see coloured dots that show where the 'water pipes' carry the water through the stems!

What you will see

The celery will suck up the water through the stalks, which are really plant stems. You will see that the coloured water has travelled up the stem. This is what happens to all plants, but we cannot usually see the water. The celery leaves will also be stained the colour of the food colouring.

Top Tip

You can do this experiment with a flower too. A white carnation left standing in strongly coloured water will turn the colour of the water!

Test your knowledge

Section 1

Alive, once alive or never alive?

Imagine you are at the beach and have found these things washed up on the sand. Can you sort them into three sets? Draw the things in the correct boxes below.

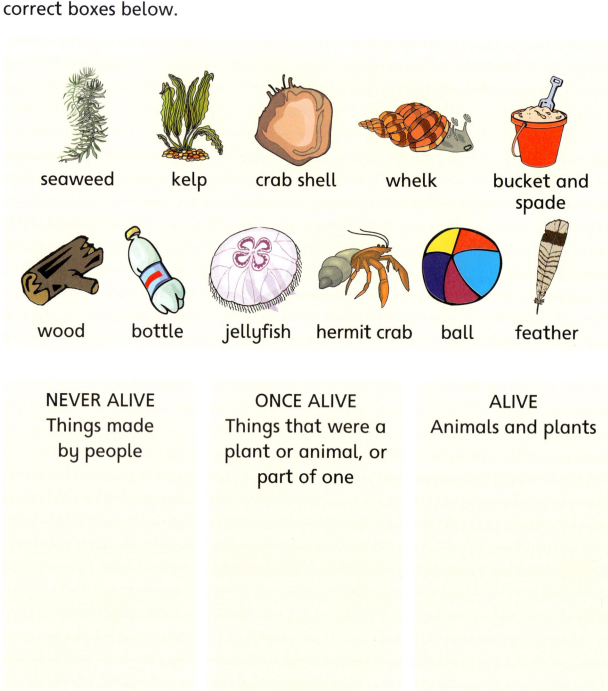

NEVER ALIVE	ONCE ALIVE	ALIVE
Things made by people	Things that were a plant or animal, or part of one	Animals and plants

Section 2

Sort it out! You could find all of these things in a rockpool. Sort them into two sets.

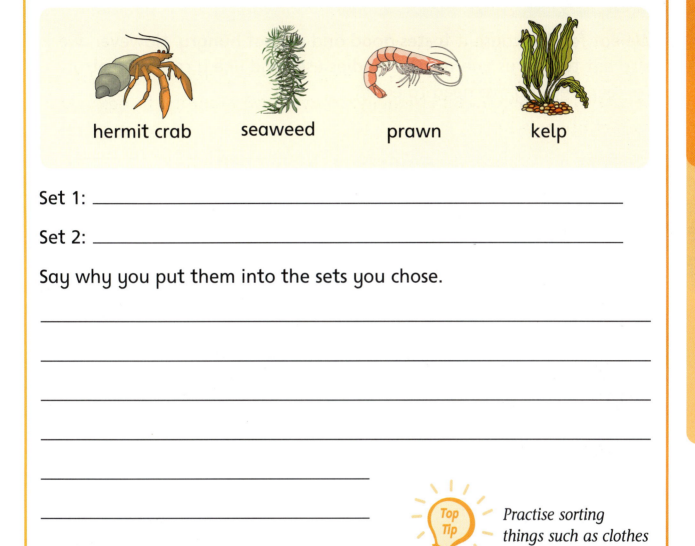

hermit crab   seaweed   prawn   kelp

Set 1: _____

Set 2: _____

Say why you put them into the sets you chose.

_____

_____

_____

_____

_____

_____

_____

Top Tip — *Practise sorting things such as clothes into sets.*

I'm alive!

Except when you sit in front of the TV – then you're a couch potato!

Eating healthy food

Why do we eat food?

We eat food because it **tastes good** and **we get hungry**. However, we also eat to **give us energy**. Our bodies use food like a car uses petrol. To keep healthy, we need to eat:

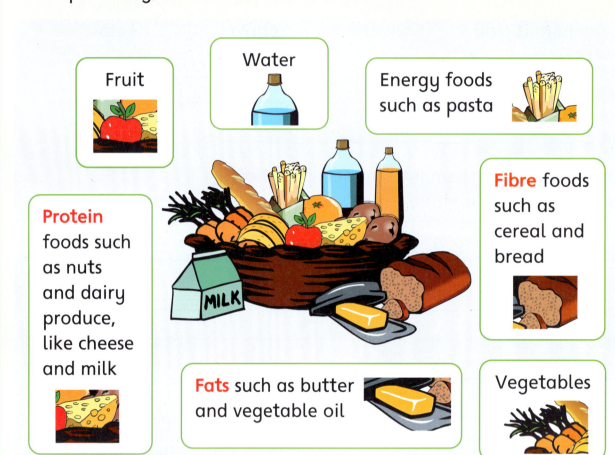

Fruit

Water

Energy foods such as pasta

Protein foods such as nuts and dairy produce, like cheese and milk

Fibre foods such as cereal and bread

Fats such as butter and vegetable oil

Vegetables

Food for different jobs

The different types of food we eat do different jobs. We need the **vitamins** in fruit and vegetables to make our bodies work properly. The protein in foods such as cheese and nuts **helps our bodies to grow** and also helps us to heal when we hurt ourselves. Fibre helps us to **digest** our food. Fats and energy foods such as pasta help us to run about and play.

Top Tip
Make sure you eat 5 portions of fruit and vegetables every day. Check out this website for fun activities: http://www.dole5aday.com/

Planning a healthy lunchbox

No food is entirely bad. However, there are some foods such as sweets and fizzy drinks that have lots of sugar in them. They are bad for your teeth and fill you up, so you do not want proper meals. Look at these two lunchboxes. They are filled with lots of healthy food.

Careful though – each box has some food in it that would not be healthy if you ate too much, such as sugary chocolate biscuits and salty, fatty crisps.

I love strawberries. I could eat them all day long!

I think you did! There are none left...

Have a go...

Plan a healthy lunchbox using some of this food:

bread, cheese, raisins, orange, apple, flapjack, apple juice, milk

Key words

protein   vitamins

fats   digest

fibre

Quick Test

Finish the sentences.

1. Vegetables are good for you because…

2. We should not eat too many sweets because…

3. We need energy foods to help us to…

4. We need protein-filled foods to…

5. Fibre helps us to…

Our bodies

Bones and muscles

Have you ever thought why you have **bones**? Because if you did not, you would be all floppy! **Bones make our bodies strong**. Without bones we could not move or stand up.

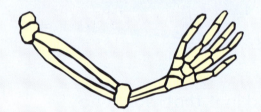

Muscles **help our bones to move**. When we bend up our arm, the muscles go tight and help to pull the bones where we want them to go.

Top Tip

If you tighten the muscles in your arms or legs, you can see their shape through your skin. If you feel them, you can feel them tighten and relax.

If we didn't have bones, we'd be like wobbly jelly!

That would make riding a bike difficult – have you ever seen a jellyfish on a bike?

Special bones

We have some very special bones.

Feel your head. Can you feel the hard bone inside? It is called a **skull**. It is a special bony box that **keeps your brain safe**.

Feel the middle of your back. Can you feel the bony bumps in a line? That is your **backbone**. Some people call it your **spine**. Your backbone helps you to **stand up straight**, but, because it is made of lots of small pieces all joined together like a row of cotton reels, it helps you to bend and stretch too!

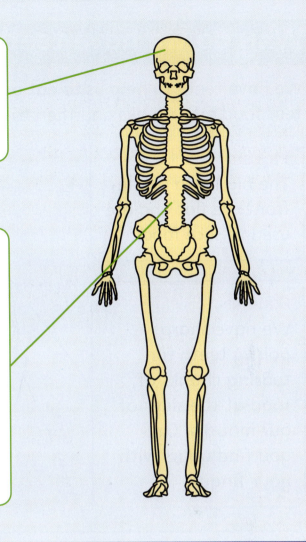

Have a go...

Bend your fingers and feel them with your other hand. Can you feel the bones and muscles moving? Try this with all the different joints in your body – knees, elbows, wrists, etc.

Key words

bones	backbone
muscles	spine
skull	

Quick Test

1 What is the bone called that keeps your brain safe?

2 Give two names for the bone in your back.
   a _____
   b _____

3 Why do we have bones?

4 Why do we have muscles?

Looking at teeth

Why do we have teeth?

We have teeth to **help us to eat our food**. We use them to bite and tear food into chunks and then to chew our food before we swallow it.

Incisors
The big teeth at the front are flat and are used to **slice** food.

Canines
We have sharp pointy teeth for **tearing** chunks of food at the sides of our mouths. Can you find yours with your finger?

Molars
We have big flat teeth for **grinding** up food at the back of our mouths. Put your finger in your mouth. Can you feel them?

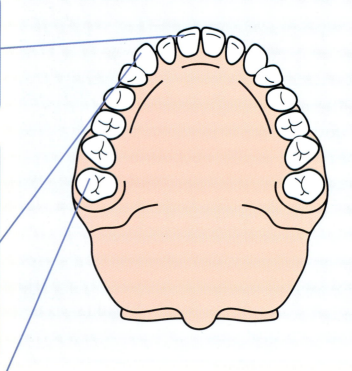

Top Tip
Listen to the radio while you clean your teeth. You should carefully clean them for the length of one whole song to get them really clean.

I get stickers when I visit my dentist. I like the special chair too!

I like spitting out the pink mouthwash!

Keeping our teeth healthy

It is very important that we keep our teeth clean. To do this, we need to clean our teeth properly twice every day.

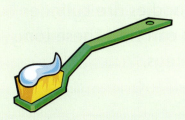

It is also important to **visit the dentist** for check-ups to make sure our teeth are healthy.

We should not eat lots of sticky, sugary food, because it sticks to our teeth. It burns away the special hard coating on our teeth called **enamel**. This is how **cavities** are formed.

Bacteria in our mouths break down the sugar and make **acid**. This acid can damage our teeth.

Have a go...

Look at your pets – or a friend's pet – as they eat. Are their teeth like yours or are they different? Do not poke your pet's mouth – they may not like it and you could get bitten!

Key words

incisors	cavities
canines	bacteria
molars	acid
enamel	

Quick Test

1. Why do we have teeth?
2. What are the big, flat teeth at the back of our mouths for?
3. Why should we clean our teeth?
4. What are incisors for?
5. What are canines for?

Test your skills

Make a bone model

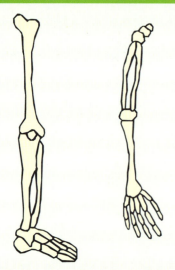

You can make models to help you to understand how bones work and why they are strong.

The long bones in our bodies are cylinder shapes, like tubes. You can find these long bones in our arms and legs.

1 Take a piece of paper. It could not support any weight, could it?

2 However, if you roll it into a tube, it is much stronger! Make a roll of paper and fasten it with tape. A tube is a strong shape.

3 If you stand it on its end, it will hold up small things such as plastic plates!

I'm going to make a whole skeleton.

That would be great for Halloween!

Make some more bones!

Bend your arm up and down. If you put your hand on your elbow you can feel the bones moving, like a **hinge**.

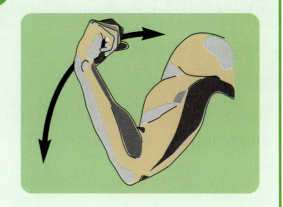

Make a hinge by cutting two pieces of card. A cereal box will do. Join the pieces together to make a hinge using tape so you can bend it up and down – just like your arm!

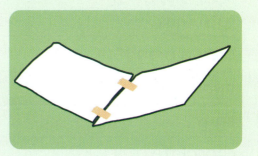

And even more bones!

Thread some cotton reels onto a piece of string. This is like a backbone.

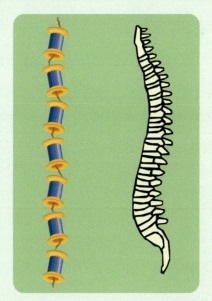

Aren't bones amazing?

Test your knowledge

Section 1

Imagine you are planning a picnic to take to the park. Tick your shopping list to show the 10 healthiest choices.

	Healthy
satsumas	
brown bread	
apples	
eggs	
chocolate cake	
crisps	
bananas	
biscuits	
carrot sticks	
cheese	
milk	
lemonade	
blackcurrant juice drink	
orange juice	

I love raisins.

Mmm...especially chocolate-coated ones!

Section 2

1   Pretend you are planning a talk about healthy eating for a school project. Match the food cards to the job cards.

fats to give us lasting energy

vitamins to help us to stay healthy

energy foods so we can play

protein foods to help us to grow

2   Put a circle round the healthier food in each pair:

orange juice
fizzy orange drink

apple
apple pie

crisps
raisins

jelly
fruit salad

Make sure you eat lots of fruit and vegetables every day!

Top Tip

Is it natural?

Materials

Can you tell if a **material** is **natural** or made by people?

It is harder than it sounds!

These are natural materials. Do you know where they come from?

wood   stone   sand

wool   leather   oil

Wood comes from trees.

Stone and sand are dug from the ground in special places called **quarries**.

Wool comes from sheep and leather is animal skin.

Oil is a liquid made from **tiny creatures that died millions of years ago,** in the time of the dinosaurs.

Who'd have thought that glass is made from sand?

And who'd have thought your head wasn't made from wood!

Made by people?

Materials made by people include glass, plastic and ceramics. In the beginning, however, they were natural materials too!

Glass is made from heated sand.

Plastic is made from oil.

Ceramics are made from clay.

'Made by people' just means that the material has been **greatly changed** by the actions of people.

 **Top Tip**

Have a look at this site to find out how oil is made into things:
http://www.priweb.org/ed/pgws/uses/uses_home.html

Have a go...

Look around your house. Which materials are natural and which are made by people?

Key words

material   natural

Quick Test

1. Which of these things are made from natural materials?

   wooden chair
   glass jar
   plastic toy car
   woollen jumper
   leather shoes

2. Match the material made by people to the natural material used to make it:

Material used:	Changed into:
sand	ceramic mug
clay	plastic doll
oil	glass window

Floating and sinking

Which things float?

Have you noticed that some things **float** on water and other things **sink**? Can you predict which things will float?

Think about the toys in your bath. What are they made from?

Do you think there are any rules about **which things float and which things sink?**

Testing materials

These things have been tested to see if they float or sink. Look at the results.

floats	sinks
cork	pebble
paper boat	coin
matchstick	marble
plastic ball	play clay
play clay	

Top Tip

Do not assume that all heavy things sink. What about big, heavy metal boats? They float because the air inside them makes them light for their size. Also, the big bottoms give the water plenty to push up against, so they float.

Light or heavy for their size?

Things float because they are light for their size. If they are heavy for their size, they sink. A stone sinks because it is heavy for its size and a cork floats because it is light for its size.

The play clay that was rolled into a ball sank, because it was heavy for its size. The play clay that was made into a boat shape, however, floated. It weighed the same as the ball, but making it bigger by squashing it into a boat shape gives the water a bigger area to push against. This makes the boat light for its size, so it floats.

It's the same with people – if you lie on your back in a swimming pool, then you'll float!

I prefer jumping in so I can get to the bottom!

Have a go...

Make a play clay boat and see how many paper clips you can put inside before it sinks. Then make the boat bigger and see if it holds more before it sinks.

Key words

float   sink

Quick Test

1 Which things will float?

cork   pebble   coin   matchstick   metal spoon   plastic spoon

2 Why did the play clay boat float and the play clay ball sink?

3 Why do some things float?

4 Why do some things sink?

31

Looking at materials

In the toybox

Your toys are made from different sorts of materials.

Why is a teddy made of fur fabric? Think about the **properties** of the material. That just means **what the material is like and what it can do**. Fur fabric is soft and warm and good to touch. A teddy is for cuddling, so fur fabric is a good material to make a teddy from.

Why is a toy car made from metal? Think about the properties of the material. **Metal is strong and tough**. A toy car needs to be strong enough to push around the floor or a track. It does not have to be soft or cuddly. Metal is strong, so it is a good material to make a toy car from.

I thought material was the stuff clothes are made from!

No, Sam – that's fabric. Materials means whatever something is made from, like the fur made into the teddy you hide under your bed!

In the wardrobe

Different fabrics are chosen for different sorts of clothes depending on the job they will have to do.

A rain jacket could be made of plastic, because it is waterproof.

A paper raincoat would not be very useful, because it would get soggy and fall apart! We need to think about the properties of the materials we choose.

A woolly jumper is **soft and warm**, so we wear it in winter. In the summer, we want cooler clothes, so we do not make swimming costumes or shorts from wool!

Top Tip

Remember, 'material' just means what something is made from. It could be glass, plastic, wood, stone – anything!

Have a go...

Look at your toys. What materials are they made from? Why are they made from those materials?

Key words

properties

1 Match the material to the toy. How did you decide?

Toys:

> cuddly rabbit   train
> ball   doll

Materials:

> plastic   rubber   wood
> fur fabric

2 Why are lunchboxes often made from plastic?

3 Why are clothes made from fabric?

Changing materials

Changing

There are lots of ways to change things. We can change the shape of certain things by **squashing, bending, twisting and stretching** them. Think of a flat, smooth piece of paper. How many ways can you change it?

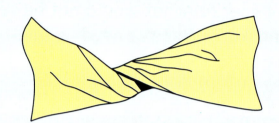

Heating and cooling

We can also change things by heating and cooling them. When **water is heated, some of it turns into** steam **as the water boils**. When chocolate is heated, it goes runny.

If we **cool the steam down it will turn back into water**. Have you noticed drops of water on a mirror in the bathroom? That happens when hot bath water turns into steam, then hits the cold mirror and turns back into water.

If you leave a bar of chocolate in the car on a sunny day, it will melt. If you put it in the fridge to cool it down, it will go hard again.

Top Tip

Heating and freezing things uses very high and very low temperatures, which can be dangerous, so always ask an adult to help you.

Things that don't change back

There are some things we can change by heating or cooling that **we cannot change back.**

Think about wood that has been burnt on a bonfire. It turns into a grey powder called ash. We cannot change ash back into wood.

I like cooling juice until it freezes – then I have an ice lolly!

And I like heating chocolate until it melts – and pouring it on my ice cream so it goes hard again!

Have a go...

Next time you eat an egg, think about how heat has changed it. Could you change it back?

Key words

steam   melt

Quick Test

1 What could you do to a piece of clay to change it?

2 Which of these things can be changed back after heating has changed them?

butter   ice   cake mixture

3 What happens when steam cools down?

4 How could you change water into steam?

5 Name a place where you can see steam change back into water.

Test your skills

Make a bubble speedboat

What you need:

- Scissors
- Play clay
- Plastic washing-up bottle (empty)
- Baking soda – this is used for making cakes
- Vinegar
- Kitchen roll paper or paper hanky
- Plastic bendy drinking straw

I'd love to drive a speedboat really fast.

You can't even control your skateboard yet!

What you do

- Ask an adult to pierce a small hole in the washing-up bottle from the outside, using a pair of scissors. The hole should be in the bottom of the bottle, near the edge, and big enough to hold a straw.

- Push the straw into the hole, leaving the bendy end sticking out. Bend the straw down slightly – this will be the boat's propeller! Press clay around the hole to make a tight fit.

- Put some baking soda (about half a cup) into some kitchen roll paper or a paper hanky. Roll it up into a long, thin sausage and twist the end to seal.

- Pour some vinegar (about a cupful) into the washing-up bottle.

- Push the tissue roll into the bottle and put the stopper back on.

- Put the bottle gently into a bath of water and watch it rush along by itself!

How it works

As the tissue paper soaks up the vinegar, the sausage shape opens. The soda mixes with the vinegar and makes foam and gas. As the gas pushes out of the bottle through the straw, it moves the boat along.

Test your knowledge

Section 1

1   Put a circle round the things that are made from plastic.

doll   book   saucepan   light bulb

2   Which of these things are natural and which are synthetic (made by people)? Put a tick below the natural ones.

flower   pebble   book   feather   leaf   bucket and spade

☐   ☐   ☐   ☐   ☐   ☐

3   Match the material to the sentence that tells us what happens when it is heated:

chocolate   ice   wood

When it is heated (burnt) it changes to a powder called ash. We cannot change it back.

When it is heated it goes runny. If we put it in the fridge it would go hard again.

When it is heated it changes into water. If we put it back in the freezer it would go hard again.

Section 2

Which material would you choose for each object? Why? Underline the most appropriate material.

OBJECT		MATERIAL
cuddly toy		fur stone
chair		rubber wood
saucepan		metal chocolate
bus		metal paper
rabbit hutch		wood fabric
spoon		plastic cardboard
shoes		leather glass

Top Tip

Look around your house and think about what materials things are made from. Remember, materials are chosen to suit what an object will be used for.

I think a rubber chair would be great fun!

And with glass shoes, I could pretend to be Cinderella!

Batteries and bulbs

Electricity at home

Electricity gives us heat, light, sound and pictures. Have you ever thought about how many things in your house use electricity? Take a look around!

Hair raising!

Have you ever taken off your jumper and heard a **crackling noise**, or seen your hair stand on end? That is caused by static electricity. Static electricity can be made by rubbing a balloon on your jumper. If you then hold it near your hair, it will stand up on end!

Ooh, Sam! Are you scared? Your hair is standing on end!

No, it's my electric personality!

Electricity can be dangerous

The electricity that runs televisions and lights – the sort we use by plugging things into sockets – can be very dangerous. We call this **mains electricity**.

Never push anything into a socket or use plugs or switches with wet hands. You could get an electric shock **which could kill you**.

The electricity that runs toys or torches comes from batteries. This is sometimes called chemical electricity. The electricity we get from batteries is quite safe, although we should not open batteries, as they contain chemicals.

Top Tip

Check out the experiment on this website – and make your own lightning!
http://www.exploratorium.edu/ science_explorer/sparker.html

Have a go...

Blow up a balloon and rub it on your jumper. Can you make your hair stand up on end? If you move the balloon about, your hair will 'dance'!

Key words

static electricity

mains electricity

Quick Test

1 Name three sorts of electricity.

a _____

b _____

c _____

2 How can electricity be dangerous?

3 Name three things that use mains electricity.

4 Name three things that use a battery.

Sound

Have you heard?

What sounds have you heard today?

If you live in the town, you may have heard **cars, buses and sirens**. In the countryside, you may have heard sheep or cows. You will have heard voices talking wherever you live!

Do you know how we hear sounds? Noises go into our ears and make our **eardrums** move. This tells our brain that we have heard a noise. It is important **never to put or poke things in our ears**, because we can hurt them easily.

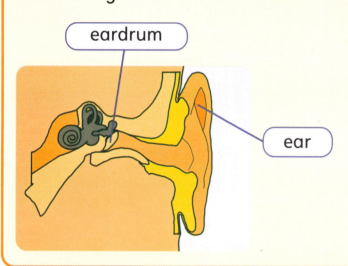

eardrum

ear

I like making a loud noise by popping balloons and making people jump!

Mum says I make a loud noise when I play my recorder!

Sound travels

Sound travels through the air in waves. We cannot see them though!

Look at the picture below. The dog is barking. She sounds very loud to the people at the bus stop because they are nearby. The girl playing football can hear the dog barking, but only quietly. This is because she is far away.

Top Tip

Go on a 'sounds walk', listening carefully. How many different sounds can you hear? You could make a 'sounds map' by drawing the things you hear and where you hear them.

Have a go...

Look at the picture above again. Who can hear the music from the boy's radio more clearly – the girl playing football or the people at the bus stop? Why?

Key words

eardrums

Quick Test

1 When noises enter our ears, what starts to move?

2 Why must we never poke things in our ears?

3 Choose the right word: Sound travels in:

   lines   waves   wiggles

4 Fill in the missing words:

   a Noises sound _____ if you are closer to the thing making a noise.

   b Noises sound _____ if you are further away from the thing making a noise.

Forces

Pushes and pulls

Forces can make things **move, stop moving, change direction, change shape and fall to the ground.**

When we talk about forces, we often mean **pushes and pulls.**

A push can make a toy car move.

A pull can make a toy move.

Squeezing a sponge uses a push as we squash the sponge with our hands.

Another force, called **gravity**, makes things fall to the ground when they are dropped. All things are **pulled towards the centre of the earth by gravity.**

Gravity must be why you always fall over!

And why you do bellyflops in our swimming class!

How to stop moving

What makes things stop moving?

When you roll a ball, it eventually stops. When you push a toy car, it will eventually stop moving.

This is because of **friction**. **Friction happens when two objects are rubbed together**. The rubbing takes away some of the **energy** of the thing that is moving and slows it down.

In the case of a toy car pushed on a table, the friction happens as the wheels rub against the table.

Rough surfaces increase friction and **smooth surfaces reduce friction**. Think of sliding in the playground. On a summer day, you would not slide at all, as the surface of the playground is rough. On an icy day you may slide, because the surface might be smooth from a coating of ice.

 **Top Tip**

You can try out another friction experiment here: http://www.smm.org/sln/tf/f/friction/friction.html

 Have a go...

Rub your dry hands together. Then add soap and water and rub them together again. With the soap and water, your hands feel slippery. This is because the soap and water reduce friction by making the surface of your hands smoother.

 **Key words**

forces	friction
gravity	energy

Quick Test

1. What force makes things fall when we drop them?

2. How does friction slow down a toy car?

3. When you squeeze a sponge, are you using a push, pull or twist?

4. Fill in the missing words:

   a Rough surfaces _____ friction.

   b Smooth surfaces _____ friction.

Sources of light

Things that make light

Some things actually **make light**, such as the sun, a candle, fire or a light bulb. We call these sources of light. Sources of light **actually glow**.

If you put them into a completely dark room, they would still glow and you would still be able to see them.

These are sources of light

There are other things that **look as though they are sources of light** because they seem to shine, such as the moon, mirrors and silver foil. These are not sources of light; they are **reflecting** light.

If you put them into a completely dark room, they would not glow and you would not be able to see them.

Top Tip
Remember, a thing is only a source of light if you would still be able to see it in a completely dark room – that is the test!

I can't see when it's dark.

Rabbits can – perhaps we should eat more carrots?

How do we see things?

We can only see things because of light! Light hits objects and bounces off them. The **light enters our eyes and we see things.**

The light hits the rabbit and bounces off. Then the light enters our eyes and we see the rabbit.

When it is dark, at night, there is much less light so it is difficult to see things. But if we turn on a light or lamp, we can see again.

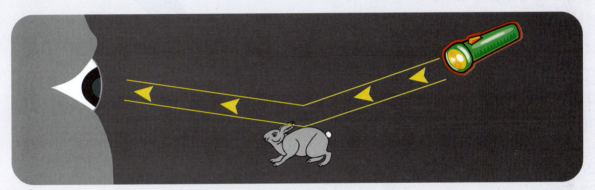

The light from the lamp hits an object and bounces off, then enters our eyes and allows us to see the object.

SOURCES OF LIGHT

PHYSICAL PROCESSES

Have a go...

Look in a magazine and cut out pictures of sources of light. Make a collage to help you to remember!

Key words

sources of light

Quick Test

Which of these things are sources of light? How do you know?

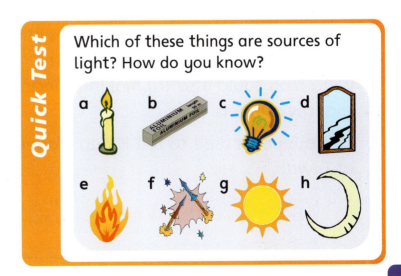

Shadows

Why do we see shadows?

Light travels in straight lines. It cannot bend. When light hits a **transparent** material, like glass, it can pass through. When it hits an **opaque** material, like wood, it cannot pass through and a shadow is made.

A **shadow shows us where the light has not passed through an object**, so the shadow is the same general shape as the object.

Your shadow is the same shape as you – but is sometimes taller or shorter than you, depending on the time of day. If you look at your shadow in the late afternoon or early evening, for example, it will be **tall and thin** because the sun is sinking low in the sky.

Top Tip

Shadows move as the sun moves across the sky. Find a shadow in the early morning. Draw around the shadow with chalk. Come back at lunchtime. Draw around it again. Come back at tea time. Has the shadow moved?

Shadow puppets

You can put on a show using shadows. Have you ever made funny shapes with your hands to make a shadow on the wall? You could also try making special shadow puppets.

This puppet is used to tell Hindu stories at Diwali.

You could make...

Try making your hands into different shapes so the shadows look like:

- A flying bat
- A dog with flapping ears
- A walking man

Make shadows on the wall or use a white sheet and a lamp as your stage.

Shadow tag is a great game!

Yes, but now I've caught your shadow, so you're it!

Have a go...

Have you ever played shadow tag? It's great fun! On a sunny day, when everyone has shadows that are easy to see, play chase. If you are 'it', you tag people by jumping onto their shadow!

Key words

transparent   opaque

Quick Test

1. Can light bend?
2. How is a shadow made?
3. What does 'transparent' mean?
4. During which part of the day would your shadow be longest?
5. What does 'opaque' mean?

Test your skills

Make jumping spiders!

What you need:

- Balloon
- Sheet of paper
- Pen
- Jumper to rub your balloon on!

balloon

jumper

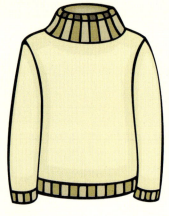

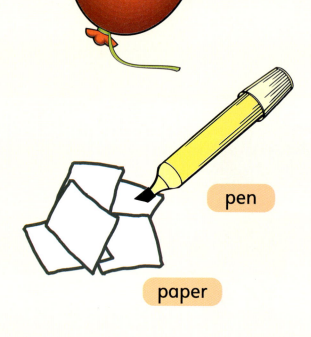

pen

paper

Yuk, I hate spiders!

But I LOVE them!

What to do

- Tear or cut a pile of tiny pieces of paper.

- Draw a spider on each piece.

- Blow up your balloon.

- Rub the balloon on your jumper and hold it just above the spiders. They will jump up and stick to the balloon!

How it works

The balloon gets charged with static electricity when you rub it on your jumper. The static electricity attracts the paper and makes the spiders jump as they stick to the balloon.

Test your knowledge

Section 1

1   Name five things that use electricity.

_____   _____

_____   _____

_____

2   Name two things you should never do with mains electricity.

_____

_____

_____

_____

_____

3   Sound travels in waves. Can you see them?

_____

4   What is the force called that makes things fall when we drop them?

_____

5   Name five sources of light.

_____

_____

_____

_____

_____

Section 2

Use the words from the box below to fill the gaps.

When you push a car, it eventually stops. This is because of something called _____. When two objects are _____ together, the rubbing takes away some of the _____ from the thing that is moving and _____ it down. In the case of a toy train pushed along the floor, the friction happens as the wheels rub against the floor.

Rough surfaces _____ friction and smooth surfaces _____ friction. On an _____ day you are more likely to slip over, because the ice will reduce the friction between your shoes and the ground.

reduce	rubbed	icy
energy	friction	slows	increase

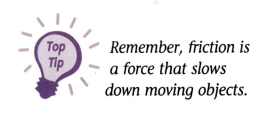

Top Tip

Remember, friction is a force that slows down moving objects.

I like sliding on the kitchen floor when mum has washed it!

That's because you're always causing friction!

National Test practice

Plants

1 **a** Fill the gaps using the words in the box.

Each part of a plant has a job. The _____ uses sunshine to make food. Plants need sunshine and water to grow. The _____ help the plant to hold itself in the soil. The _____ smells good and has a bright colour to make insects visit. Insects help the plant to make _____. The _____ is like a big pipe that takes water and goodness to all parts of the plant.

> roots   leaf   flower   stem   seeds

b Match the labels to the parts of the flower.

| roots – help to hold the plant in the soil ☐ |

| flower – attracts insects with a bright colour and lovely smell ☐ |

| stem – like a pipe that takes water and goodness to all the parts of the plant ☐ |

| leaf – uses energy from the sun to make food ☐ |

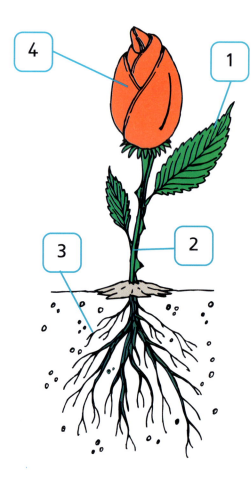

Animals and humans

2 a Match the sense to the correct part of the body.

Taste	See	Smell	Hear	Touch

with our nose	with our ears	with our fingers	with our eyes	with our tongue

b Put the animals into sets.

rabbit   cat   dog   budgie

tortoise   goldfish   snake   hamster

Has fur:	Does not have fur:

Can you think of another way of sorting the animals into sets?

_____

_____

_____

c Tick the right answer.

We have bones in our bodies:

☐ to make us look nice

☐ to stop us from being floppy and to protect parts of our body like our brain, heart, lungs

Our skull is:

☐ a bony box to protect our brain

☐ another name for our backbone

Our spine is:

☐ another name for our backbone

☐ a long bone in our leg

d Use the words in the box to fill the gaps.

Why do we have teeth?

We have teeth to help us to _____ . We use them to _____ food into chunks and then we use them to chew our food before we swallow. We have big flat teeth for _____ up food at the back of our mouths. We have _____ pointy teeth for tearing chunks of food at the sides of our mouths. The big teeth at the front are _____ and are used to slice food.

| eat our food | bite and tear | flat |
| grinding | sharp |

Materials

3 a Match the object to the best material.

OBJECT	MATERIAL
teddy	metal
dustbin bag	wood
book	glass
baking tray	plastic
window	fur fabric
door	paper

b Natural (found in nature) or synthetic (made by people)?
Look at these things and decide!

	natural	synthetic
plastic chair		
rock		
plastic toy train		
feather		
glass window		
shell		

c Wood comes from cut-down trees. Which of these things are
made using the natural material wood?

☐ newspaper ☐ plastic doll ☐ door

☐ saucepan ☐ television

Light and power

4 **a** Which of the things in the picture use electricity? There are eight. Can you find them all?

_____   _____

_____   _____

_____   _____

_____   _____

b Which of the things in the picture below are sources of light?

_____   _____

_____   _____

_____

Plants and animals

PAGES 4–5 PLANTS

1 makes food
2 to hold them in the soil
3 it draws up liquid (from the soil) and takes water to all parts of a plant
4 to attract insects
5 sunshine and water

PAGES 6–7 THE FIVE SENSES

1 smell, taste, touch, hearing, sight
2 taste buds
3 because your nose is blocked and you cannot smell them
4 special tubes
5 they help us to find things out about the world and how things work

PAGES 8–9 RECYCLING AND REUSING

1 rubbish that has been dropped on the ground, such as old bags, packets and bottles
2 animals can cut themselves
3 use things again such as plastic bags, plastic containers and glass bottles
4 animals can get stuck in them
5 recycling

PAGES 10–11 WHEN THINGS ARE ALIVE

dandelion, tiger, oak tree and guinea pig; because they feed, move, feel things, breathe, get rid of waste, grow and change, and reproduce (have offspring)

PAGES 12–13 SORTING

Many answers are possible, for example: sparrow and owl (birds), butterfly and mouse (not birds) OR: sparrow, owl and butterfly (can fly), mouse (cannot fly)

PAGES 16–17 TEST YOUR KNOWLEDGE

Section 1

Never alive: bucket and spade, bottle, ball
Once alive: crab shell, wood, feather
Alive: seaweed, kelp, whelk, jellyfish, hermit crab

Section 2

could be sorted into two groups according to plants and animals:
Plants: seaweed, kelp
Animals: hermit crab, prawn
Note for parents: could also be sorted according to other criteria, but you need to be able to say why, e.g. 'has shell', 'does not have shell', colour, etc.

Humans

PAGES 18–19 EATING HEALTHY FOOD

1 … they contain vitamins.
2 … they contain lots of sugar, which is bad for our teeth and fills us up, so we do not want other food which we need.
3 … run about and play.
4 … help us to grow and to heal any cuts or bruises.
5 … digest our food.

PAGES 20–21 OUR BODIES

1 skull
2 backbone, spine
3 to make our bodies strong and to help us to move or stand up
4 to help our bones to move

PAGES 22–23 LOOKING AT TEETH

1 to help us to eat our food
2 grinding up food

3 to stop food sticking to our teeth and attracting bacteria that break down sugar and make acid

4 slicing food

5 tearing food

PAGES 26–27 TEST YOUR KNOWLEDGE

Section 1

Can include less healthy options as long as they are balanced by healthier options, e.g. can include crisps if options such as fruit, cheese, milk, brown bread, etc. are also included.

Section 2

1  energy foods so we can play

 protein foods to help us to grow

 fats to give us lasting energy

 vitamins to help us stay healthy

2 circled words are: orange juice, apple, raisins, fruit salad

Materials

PAGES 28–29 IS IT NATURAL

1 wooden chair, woollen jumper, leather shoes

2 sand ➔ glass window
clay ➔ ceramic mug
oil ➔ plastic doll

PAGES 30–31 FLOATING AND SINKING

1 cork, matchstick, plastic spoon

2 because the boat gives the water a larger area to push against; it is light for its size, so it floats

3 because they are light for their size

4 because they are heavy for their size

PAGES 32–33 LOOKING AT MATERIALS

1 cuddly rabbit ➔ fur fabric
train ➔ wood
ball ➔ rubber
doll ➔ plastic

2 because it is hard to break, airtight, reusable and waterproof

3 because fabric is flexible and soft, making it comfortable to wear and easy to cut into shapes and sew

PAGES 34–35 CHANGING MATERIALS

1 mould it into a shape; heat it in a special oven

2 butter, ice

3 it changes (back) into water

4 by heating it until it boils

5 any sensible answers – could be bathroom mirror, window, etc.

PAGES 38–39 TEST YOUR KNOWLEDGE

Section 1

1 circled items are: doll

2 ticked items are: flower, pebble, feather, leaf

3 chocolate ➔ When it is heated it goes runny. If we put it in the fridge it would go hard again.

ice ➔ When it is heated it changes into water. If we put it in the freezer it would go hard again.

wood ➔ When it is heated (burnt) it changes to a powder called ash. We cannot change it back.

Section 2

cuddly toy ➔ <u>fur</u>, because it is soft and warm to touch. Stone would be hard and you could not cuddle it easily!

chair ➔ <u>wood</u>, because you need something strong and stiff to take your weight. Rubber would squash when you sat on it!

saucepan ➔ <u>metal</u>, because it is a good conductor of heat. Chocolate would melt!

bus ➔ <u>metal</u>, because it is strong and can support weight. Paper would not be strong enough.

rabbit hutch ➔ <u>wood</u>, because it is strong and would make a safe home for the rabbit. The rabbit could chew through fabric and escape.

spoon ➔ <u>plastic</u>, because it can be moulded into lots of shapes. Cardboard would go soggy when you used it!

shoes ➔ <u>leather</u>, because it is soft and comfortable. Glass would be no good because it could break and cut your foot!

Physical processes

PAGES 40–41 BATTERIES AND BULBS

1 static, mains, chemical
2 it can give you an electric shock and kill you
3 any examples such as: television, light bulb, microwave oven, etc.
4 any examples such as: toy car, TV remote control, torch, car, etc.

PAGES 42–43 SOUND

1 our eardrums
2 because we could damage our eardrums
3 waves
4 a louder
  b quieter

PAGES 44–45 FORCES

1 gravity
2 the wheels rub against the table or floor, and the rubbing takes away some of the car's energy
3 push
4 a increase
  b reduce

PAGES 46–47 SOURCES OF LIGHT

1 a, c, e, f, g
  because you would be able to see them in a completely dark room

PAGES 48–49 SHADOWS

1 no
2 when light hits an opaque material, that it cannot pass through
3 a material that light can pass through
4 late afternoon/early evening
5 a material that light cannot pass through

PAGES 52–53 TEST YOUR KNOWLEDGE

Section 1

1 Many answers are possible, for example: television, cooker, microwave, radio, CD player, computer, radio-controlled car, or other electrical appliances
2 never use switches/plugs with wet hands or poke things into sockets
3 no – they are invisible
4 gravity
5 light bulb, candle, sun, torch, fire, lamp, or any other light sources

Section 2

When you push a car, it eventually stops. This is because of something called **friction**. When two objects are **rubbed** together, the rubbing takes away some of the **energy** from the thing that is moving and **slows** it down. In the case of a toy train pushed along the floor, the friction happens as the wheels rub against the floor.

Rough surfaces **increase** friction and smooth surfaces **reduce** friction. On an **icy** day you are more likely to slip over, because the ice will reduce the friction between your shoes and the ground.

National Test practice

PAGES 54–55 NATIONAL TEST PRACTICE

1 Plants

a Each part of a plant has a job. The **leaf** uses sunshine to make food. Plants need sunshine and water to grow. The **roots** help the plant to hold itself in the soil. The **flower** smells good and has a bright colour to make insects visit. Insects help the plant to make **seeds**. The **stem** is like a big pipe that takes water and goodness to all parts of the plant.

b **1** leaf – uses energy from the sun to make food

2 stem – like a pipe that takes water and goodness to all the parts of the plant

3 roots – help to hold the plant in the soil

4 flower – attracts insects with a bright colour and lovely smell

2 Animals and humans

a Taste → with our tongue

See → with our eyes

Smell → with our nose

Hear → with our ears

Touch → with our fingers

b Has fur: rabbit, cat, dog, hamster

Does not have fur: budgie, tortoise, goldfish, snake

Could also be sorted into:

'Has feathers': budgie, and 'does not have feathers': rabbit, cat, dog, hamster, tortoise, goldfish, snake

OR

'Has shell': tortoise, and 'Does not have shell': rabbit, cat, dog, budgie, goldfish, snake, hamster

OR

'Lives in water': goldfish, and 'Does not live in water': rabbit, cat, dog, budgie, tortoise, snake, hamster

c We have bones in our bodies: to stop us from being floppy and to protect parts of our body like our brain, heart, lungs

Our skull is: a bony box to protect our brain

Our spine is: another name for our backbone

d Why do we have teeth?

We have teeth to help us to **eat our food**. We use them to **bite and tear** food into chunks and then we use them to chew our food before we swallow. We have big flat teeth for **grinding** up food at the back of our mouths. We have **sharp** pointy teeth for tearing chunks of food at the sides of our mouths. The big teeth at the front are **flat** and are used to slice food.

3 Materials

a teddy → fur fabric
dustbin bag → plastic
book → paper
baking tray → metal
window → glass
door → wood

b natural: rock, feather, shell

synthetic: plastic chair, plastic toy train, glass window

c door, newspaper (paper is made from wood pulp)

4 Light and power

a TV, radio, CD player/music centre, light bulb hanging from ceiling, table lamp, computer, radio-controlled car, watch

b candle, light bulb, fire, firework, candle lantern

acid a sour liquid such as lemon juice; some acids, like the ones found in batteries, can burn your skin

alive something that breathes, moves, feeds, makes waste, grows and changes, feels things and makes babies

backbone the large bone that goes down the centre of the back, sometimes called the spine

bacteria tiny living things; can be harmful to people (by causing illnesses such as tummy upsets) or helpful (turn milk into yoghurt)

bones the hard parts inside the body that support us; without bones we would be very floppy and unable to stand up

canines sharp teeth for tearing

cavities holes that form in teeth where they have decayed

digest food is digested; it is chewed and swallowed and broken down by the acids in the stomach

eardrum a bit inside the ear that helps you hear

enamel hard coating on the outside of teeth

energy what is used to make things happen, such as moving and growing. There are lots of different types of energy, such as electrical energy and light energy from the sun (solar energy)

fats fats such as butter and vegetable oil in our diets keep us healthy

fibre found in fruit, vegetables and cereals; helps us to digest our food

float to bob about on the surface of a liquid

flower the part of a plant that attracts insects; often smells nice and is brightly coloured

forces such as gravity, friction, etc.

friction happens when two objects rub together; takes away energy from moving objects and slows them down

gravity the force pulling things to the centre of the earth

incisors front teeth for cropping

insects tiny creatures, such as beetles, ants, flys, etc.

leaves where food for plants is made using energy from the sun

litter rubbish such as packaging that people throw away

mains electricity the electricity we use in our homes to run items such as televisions and cookers; can be dangerous, so you should never push anything into a plug socket

material what things are made of, such as plastic or wood

melt when ice changes back to water as it gets warmer we say it has melted

molars flat back teeth for grinding

muscles the fibres attached to our bones which help us to move as they expand and contract

natural (material) something that you could find in the environment, such as pebbles, sand, wood from trees, etc.

opaque things that light cannot pass through, e.g. a stone

properties what something is like and what it can do

protein something that is needed in our diets; it is found in foods such as nuts, meat and cheese

recycled processing materials and using them again

reuse using something again instead of throwing it away

root the part of a flower that holds it in the ground

senses sight, hearing, smell, taste, touch

sets groups made when things are sorted

sink to fall to the bottom of a liquid

skull the bony box that protects our brain

sources of light things that actually make light, such as the sun or a candle

spine another name for the backbone

static electricity an example is the electricity made when you rub a balloon on your jumper; lightning is another example

steam water vapour made when water boils

stem the part of a flowering plant that holds the flowers up in the air

taste buds tiny bumps on your tongue that help you to taste whether things are sweet or sour, for example

transparent things that we can see through, such as glass; light can travel through transparent materials

vitamins a sort of special goodness that is mainly found in food. We need lots of vitamins to keep us healthy. Vitamin C, for example, from fruits such as oranges, helps us to fight disease